KENNY SAGE

THE BIRTH & RISE OF TECHNOLOGICAL SINGULARITY

TABLE OF CONTENTS

THE BIRTH & RISE OF TECHNOLOGICAL SINGULARITY	2
INTRODUCTION	5
CHAPTER 1	12
TECHNOLOGY	12
CHAPTER 2	21
HISTORY OF TECHNOLOGICAL DEVELOPMENT	21
CHAPTER 3	31
TECHNOLOGY IN THE NEW ERA	31
CHAPTER 4	72
THE SOCIAL TECHNOLOGICAL LANDSCAPE	72
CHAPTER 5	83
TECHNOLOGY - MEANING AND EFFECTS ON LIFE	83
CHAPTER 6	100
HOW MUCH IS "INFORMATION TECHNOLOGY DEBT" HURTING YOUR BOTTOM-LINE?	100
CHAPTER 7	111
UTILIZING TECHNOLOGY TO IMPROVE PROFITS	111
CHAPTER 8	123

HOW LIFE WOULD BE AFFECTED IF TECHNOLOGY WAS TAKEN AWAY 123

CONCLUSION 133

INTRODUCTION

As I observe humanity, moving in ever-faster and faster paces, racing to achieve ever more at dizzying speeds even as technology threatens to outpace our very mental capacity, something is amiss. Some deeply held idea that we seem bent on fulfilling, a frantic technologically powered promise, has been broken.

Whether we realize it or not, underneath all this amazing technology we are creating, is a subtle but powerful promise: that we can accomplish more, in less time, and thereby achieve a greater quality of life.

Oh, at first the idea is seductive. Let's build a machine that can do the work in half the time! We can work in the morning and play in the afternoon. This works great in theory, except it is rarely practiced. No, once that amazing whiz-bang machine is built, it's run 24/7, working employees to the bone, so we can produce a gazillion times more in a fraction of the time! By all rights there should be a lot more people loafing. Or at least, having a high □uality of life. But are they?

How is it that our very lives are powered by machines that admittedly double in speed every 2-years, yet as a nation we are poorer than ever, more tired than ever, and less able to enjoy life as we know it? Who doesn't walk around with more lines on their foreheads even as the world races by? Whose stress levels are lower thanks to the amazing advances in technology? I don't know many.

Ladies and gentlemen, there's a conspiracy afoot. Yes, really. As a society, our job is to care about each other and improve our quality of life personally and collectively, yet the very technology that has promised to provide this is doing just the opposite. In fact it's aggregating wealth into fewer and fewer hands, and in a very real sense oppressing the rest, creating a new kind of upper class, a "technorati" if you will, that is able to harness technology to their advantage. And despite all the nifty perks of technology, are our lives really better?

Sure, we can point to increased efficiencies. Information can be transferred faster and in larger ☐uantities than ever before, and computers can crunch numbers in ever-larger chunks.

Yet have we ever stopped to ask, is that always necessarily good? Computers enable people to make mistakes, faster. Think about that for a moment.

And besides simple "business gains," and increased production, what are the actual tangible gains in human terms? Are employees happier, or are they working just as many hours as in 1960?

And another important measure: do people feel more connected to one another, with all the gizmos for interaction?

Ironically, technology tends to isolate people rather than bring them together. It promotes anonymity, and separation by encouraging us to interface over longer and longer distances, using bits of metal and plastic for the interactions. What happened to the warmth of a handshake? Looking someone in the eye? Something is getting lost in the digital revolution, and it's in the intangible, and arguably more important, realm of our lives.

What about all the fancy speed of the technology; surely this is making the world more efficient, right?

Can anyone point to studies showing the increased production and sheer extra volume of goods, services, and foods, are actually being circulated to those in need? Is the human family as a whole benefitting from the excess, or is the wealth being concentrated by those in position to take advantage of the windfall?

Again, this answer is obvious; the human family as a whole is not reaping the benefits of the technological advances, as evidenced by similar or worse levels of poverty, literacy, living conditions, and general conditions of peoples throughout the earth. Sure, there are certain segments of the population that are benefitting, yet we see the makings of a "digital divide" in which the middle classes are disappearing, while the ranks of upper and lower classes continue to swell, in large part due to technology which aggregates more and more power into the hands of those at the top. This has always been a classic harbinger of trouble, for those that care to pay attention.

And don't get me started on our amazing scientific advances in healthcare; what passes for healthcare, rightly should be labelled "sick-care" as it uses 2 main modes of operation: cutting and drugging. The human body does not generally suffer from lack of cuts, or chemicals. Many of the greatest bits of wisdom from thousands of years of human survival are being summarily censored, and even outlawed by those in charge. Don't believe me? Tsk tsk. Google it. As the saying goes, "just because you are paranoid doesn't mean they aren't after you."

Then, perhaps, is the planet better off for all the amazing increased production technology affords? Again, ha! No, the planet is hurting, possibly getting a temperature, and for sure getting filled valley to mountain with more trash than ever before.

What about the potential of the human brain: surely we are getting smarter and smarter every year, and children are benefiting greatly from exposure to all this marvelous new stuff, right?

Well, er, sort of. Actually what I've found is that we think differently, not necessarily better, than before. We have greater capacity to multi-task because, surprise, we're constantly bombarded with the need to process so much at once. But this comes at the expense of the ability to really concentrate. I'm not sure that being "scattered" is better than being "focussed."

What about social skills? Are computers and technology enhancing these? Evidence indicates that our new silicon taskmasters don't have hearts, and our capacity for human understanding, compassion, and love are not enhanced to any significant degree by our technology.

I could go on and on. The basic, simple question is, what are the actual real benefits of this technological craze we are in? Are we honestly better off with the technology than before? And what is driving this insane rapid-fire chase, and what is it for? Or have we lost site of the goal we are running towards? If it's the betterment of mankind, we are off-course.

One can always argue that the technology is not the problem; it's the biology operating it, and this is a truthful point. Technology is neither good nor bad but in the hands of its users. Truly we can do amazing things with our new gizmos, but will we? Are we? Currently, generally not.

So the jury is out. Thanks to the crazy advances in computers and technology, we can do more than ever, but the results are that we as a species are not better off in tangible ways, en masse.

In short, technology is not making the world, the planet, or people as a whole, better off; in fact many are worse off and we have the makings of a technorati that control the rest; how is this different than those in power making the world better for themselves at any other time in history?

With great power comes great responsibility. 'Nuff said.

CHAPTER 1

TECHNOLOGY

What is technology? I bet that everyone you ask will give you a different definition, depending on the things he/she uses. I'm sure your mother will tell you that technology makes her life easy because of all the gadgets that were invented, your younger brother will say technology means latest generation computers and high detailed games, you will probably say that it is your mobile phone. I will say for me technology is semi-cocked food. It was probably the best invention ever. But it's just me.

So, we have established how we, the mortals, see technology. But wouldn't be really interesting to see what it really is? Well it deals with how a species customs and knowledge of instruments meant to ease the life and how it affects its capability to fit into the environment. This is a general definition.

We, the humans see it as a result of the interaction of different sciences and engineering.

Technology is one of those terms that actually cannot be defined. It can mean everything from a main board to a way of organizing a factory. It can refer to practically everything, because all of the things around us were, at a certain point technological breakdowns.

How can we use it? As such, in specific areas like "medical technology", describing only one aspect of science or in expressions like "state-of-the-art-technology", which is by far more abstract.

We have used technology for as far as we existed. It is in our blood to discover all kinds of things meant to ease our life. And if they do not exist, you can bet wee will invent them. Starting with the discovery of fire and with the adaptation of all the things we find in nature to our needs and ending with the World Wide Web and the space satellites launched into the orbit, we managed to turn everything in our favor. And that is technology. Well, as our ancestors used to say, since the invention of the wheel we've learned a great deal about controlling our environment.

Technology is every ware around us, permitting us to interact at a global scale. Imagine that! With just a click you can see your friend from the other side of the world in real time. Not just talk to her, but see her, see what she does and how she reacts.

But technology brought upon us a great curse too. We didn't invent only good things. There are a lot of destructive gadgets that do not ease our existence, but only complicate it, or end it. Bang-Bang! And that's not your kid's gun, but a real one that can end your life in a heart beat.

All in all technology is what makes the world go round every day. Well, not exactly. Magic, or the laws of nature make that. Technology is everything else, all the things that can't be blamed on magic. Ant that's about 99,9% of the things that you use in life. So, VIVA Technology.

THE IMPORTANCE OF TECHNOLOGY

Technology refers to the collection of tools that make it easier to use, create, manage and exchange information.

In the earlier times, the use of tools by human beings was for the process of discovery and evolution. Tools remained the same for a long time in the earlier part of the history of mankind but it was also the complex human behaviors and tools of this era that modern language began as believed by many archeologists.

Technology refers the knowledge and utilization of tools, techniques and systems in order to serve a bigger purpose like solving problems or making life easier and better. Its significance on humans is tremendous because technology helps them adapt to the environment. The development of high technology including computer technology's Internet and the telephone has helped conquer communication barriers and bridge the gap between people all over the world. While there are advantages to constant evolution of technology, their evolution has also seen the increase of its destructive power as apparent in the creation of weapons of all kinds.

In a broader sense, technology affects societies in the development of advanced economies, making life more convenient to more people that have access to such technology. But while it continues to offer better means to man's day to day living, it also has unwanted results such as pollution, depletion of natural resources to the great disadvantage of the planet. Its influence on society can also be seen in how people use technology and its ethical significance in the society. Debates on the advantages and disadvantages of technology constantly arise questioning the impact of technology on the improvement or worsening of human condition. Some movements have even risen to criticize its harmful effects on the environment and its ways of alienating people. Still, there are others that view technology as beneficial to progress and the human condition. In fact, technology has evolved to serve not just human beings but also other members of the animal species as well.

Technology is often seen as a consequence of science and engineering. Through the years, new technologies and methods have been developed through research and development. The advancements of both science and technology have resulted to incremental development and disruptive technology. An example of incremental development is the gradual replacement of compact discs with DVD. While disruptive developments are automobiles replacing horse carriages. The evolution of technologies marks the significant development of other technologies in different fields, like nano technology, biotechnology, robotics, cognitive science, artificial intelligence and information technology.

The rise of technologies is a result of present day innovations in the varied fields of technology. Some of these technologies combine power to achieve the same goals. This is referred to as converging technologies. Convergence is the process of combining separate technologies and merging resources to be more interactive and user friendly. An example of this would be high technology with telephony features as well as data productivity and video combined features. Today technical innovations representing progressive developments are emerging to make use of technology's competitive advantage. Through convergence of technologies, different fields combine together to produce similar goals.

A GUIDE OF INFORMATION TECHNOLOGY

Information Technology department consisted of a single Computer Operator, who might be storing data on magnetic tape, and then putting it in a box down in the basement somewhere. The history of information technology is fascinating. Information technology is driven by the demands of the new, competitive business environment on the one hand and profound changes in the nature of computers in the other. IT systems come in the shape of many technologically advanced devices which help deliver important to managers who in turn, use this information to make crucial decision regarding the operations of their organization.

Information or Internet Technology can come in the form of computers, robots, sensors, and decision support systems. The newest form of Information & Technologies which is being implemented on the market today is the use of handholds to aid managers and subordinates in their day to day operations. Computer Technology Auditing (IT auditing) began as Electronic Data Process (EDP) Auditing and developed largely as a result of the rise in technology in accounting systems, the need for IT control and the impact of computers on the ability to perform attestation services.

IT is revolutionizing how business operates. Advanced technologies is becoming the single-most powerful force shaping the structure and functioning of work organizations, plants, offices, and executive suites. When people hear the words "Information & business technology," the first thing that comes to their mind are computers and the Internet. It may also bring up words like "network," "intranet," "server," "firewall," "security," as well as more arcane expressions such as "router," "T-1," "Ethernet," or the mysterious and exotic-sounding "VOIP." the term "Business IT" is not new and has not always referred to things relating to computer. Information technology is as old as the brain itself, if you think of the brain as an information processor. As far as IT being a science, even that goes back as far as the earliest attempts to communicate and store data. And that is essentially what information technology is: the communication and storage of information, along with the ability to process and make use of the information stored.

Information technologies is the use of computers and software to manage information. In some companies, this is referred to as Management Information Services (or MIS) or simply as Information Services (or IS). The information technology department of a large company would be responsible for storing information, protecting data, processing the techniques, transmitting the information as necessary, and later retrieving information as necessary. The benefits resulting from IT Technology benefits were in many different forms such as allowing all firms to accomplish tasks they had been doing before at reduced costs, it opened up possibilities to do things never attempted before, also allowed firms to "re-engineer" parts of their companies and make better strategic positions.

CHAPTER 2

HISTORY OF TECHNOLOGICAL DEVELOPMENT

Nineteenth century developments such as the telegraph and telephone marked the beginning of the rapid growth in information technology. All method of communication beginning with the development of language itself can be considered technological developments. However, the inventions toward the end of the nineteenth century and the beginning of the twentieth, such as the telegraph and the telephone, marked the beginning of the rapid growth leading to today's every-changing information technology. Telex machines, the direct ancestor of E-mail, are not used much anymore. Faxes are widely used, and their use increases daily. Faxes are also prosecutors of today's E-mail and networking systems.

In the 1960s, some companies became attached to computer technology to handle data processing. The computers used by these progressive companies were huge mainframes, with tubes and reels of storage tape; they were so big than they often filled a large room/ terminals_videos screens with keyboards_were hooked up to the mainframe.programming had to be done from scratch because there was no packaged software, and computer programmers, often people with no experience in business or management, owned the technology. Development in information technology that led to more powerful and less expensive personal computers have facilitate the growth of electronics information in today's business.

By the 1970s, more people had computer terminals that had access to central information on large mainframes. Some packaged software was developed so that certain tasks did not have to be programmed from scratch.however, computers were expensive, and costs rose as companies without clear needs for them were persuaded to invest in information technology.

The transformation of telecommunications in the 1980s, with the development of fiber optics, local area networks, and satellite technology, along with new powerful personal computers, facilitated the growth of information technology in organizations. Organizations now have laptop computers, desktop publishing capabilities, electronic spreadsheets, and word processing programs to gather, store, and communicate information. Turmoil and change are the norm of information technology, and they reflect and influence the concurrent changes in the business organizations on structure, profit, people, and society.

WHAT IS THE RELEVANCE OF TECHNOLOGY?

"Technology in the long-run is irrelevant". That is what a customer of mine told me when I made a presentation to him about a new product. I had been talking about the product's features and benefits and listed "state-of-the-art technology" or something to that effect, as one of them. That is when he made his statement. I realized later that he was correct, at least within the context of how I used "Technology" in my presentation. But I began thinking about whether he could be right in other contexts as well.

Technology is the making, modification, usage, and knowledge of tools, machines, techniques, crafts, systems, and methods of organization, in order to solve a problem, improve a preexisting solution to a problem, achieve a goal, handle an applied input/output relation or perform a specific function. It can also refer to the collection of such tools, including machinery, modifications, arrangements and procedures. Technologies significantly affect human as well as other animal species' ability to control and adapt to their natural environments. The term can either be applied generally or to specific areas: examples include construction technology, medical technology, and information technology.

TECHNOLOGY IS AN ENABLER

Many people mistakenly believe it is technology which drives innovation. Yet from the definitions above, that is clearly not the case. It is opportunity which defines innovation and technology which enables innovation. Think of the classic "Build a better mousetrap" example taught in most business schools.

You might have the technology to build a better mousetrap, but if you have no mice or the old mousetrap works well, there is no opportunity and then the technology to build a better one becomes irrelevant. On the other hand, if you are overrun with mice then the opportunity exists to innovate a product using your technology.

Another example, one with which I am intimately familiar, are consumer electronics startup companies. I've been associated with both those that succeeded and those that failed. Each possessed unique leading edge technologies. The difference was opportunity. Those that failed could not find the opportunity to develop a meaningful innovation using their technology. In fact to survive, these companies had to morph oftentimes into something totally different and if they were lucky they could take advantage of derivatives of their original technology. More often than not, the original technology wound up in the scrap heap. Technology, thus, is an enabler whose ultimate value proposition is to make improvements to our lives. In order to be relevant, it needs to be used to create innovations that are driven by opportunity.

TECHNOLOGY AS A COMPETITIVE ADVANTAGE?

Many companies list a technology as one of their competitive advantages. Is this valid? In some cases yes, but In most cases no.

Technology develops along two paths - an evolutionary path and a revolutionary path.

A revolutionary technology is one which enables new industries or enables solutions to problems that were previously not possible. Semiconductor technology is a good example. Not only did it spawn new industries and products, but it spawned other revolutionary technologies - transistor technology, integrated circuit technology, microprocessor technology. All which provide many of the products and services we consume today. But is semiconductor technology a competitive advantage? Looking at the number of semiconductor companies that exist today (with new ones forming every day), I'd say not. How about microprocessor technology? Again, no. Lots of microprocessor companies out there. How about Quad core microprocessor technology? Not as many companies, but you have Intel, AMD, ARM, and a host of companies building custom Quad core processors (Apple, Samsung, Qualcomm, etc). So again, not much of a competitive advantage. Competition from competing technologies and easy access to IP mitigates the perceived competitive advantage of any particular technology. Android vs iOS is a good example of how this works. Both operating systems are derivatives of UNIX. Apple used their technology to introduce iOS and gained an early market advantage. However, Google, utilizing their variant of Unix (a competing technology), caught up relatively quickly. The reasons for this lie not in the underlying technology, but in how the products made possible by those technologies were brought to market (free vs. walled garden, etc.) and the differences in the strategic visions of each company.

Evolutionary technology is one which incrementally builds upon the base revolutionary technology. But by it's very nature, the incremental change is easier for a competitor to match or leapfrog. Take for example wireless cellphone technology. Company V introduced 4G products prior to Company A and while it may have had a short term advantage, as soon as Company A introduced their 4G products, the advantage due to technology disappeared. The consumer went back to choosing Company A or Company V based on price, service, coverage, whatever, but not based on technology. Thus technology might have been relevant in the short term, but in the long term, became irrelevant.

In today's world, technologies tend to quickly become commoditized, and within any particular technology lies the seeds of its own death.

TECHNOLOGY'S RELEVANCE

This article was written from the prospective of an end customer. From a developer/designer standpoint things get murkier. The further one is removed from the technology, the less relevant it becomes. To a developer, the technology can look like a product. An enabling product, but a product nonetheless, and thus it is highly relevant. Bose uses a proprietary signal processing technology to enable products that meet a set of market requirements and thus the technology and what it enables is relevant to them. Their customers are more concerned with how it sounds, what's the price, what's the quality, etc., and not so much with how it is achieved, thus the technology used is much less relevant to them.

Recently, I was involved in a discussion on Google+ about the new Motorola X phone. A lot of the people on those posts slammed the phone for various reasons -

price, locked boot loader, etc. There were also plenty of knocks on the fact that it didn't have a quad-core processor like the S4 or HTC One which were priced similarly. What they failed to grasp is that whether the manufacturer used 1, 2, 4, or 8 cores in the end makes no difference as long as the phone can deliver a competitive (or even best of class) feature set, functionality, price, and user experience. The iPhone is one of the most successful phones ever produced, and yet it runs on a dual-core processor. It still delivers one of the best user experiences on the market. The features that are enabled by the technology are what are relevant to the consumer, not the technology itself.

The relevance of technology therefore, is as an enabler, not as a product feature or a competitive advantage, or any myriad of other things - an enabler. Looking at the Android operating system, it is an impressive piece of software technology, and yet Google gives it away. Why? Because standalone, it does nothing for Google. Giving it away allows other companies to use their expertise to build products and services which then act as enablers for Google's products and services. To Google, that's where the real value is.

The possession of or access to a technology is only important for what it enables you to do - create innovations which solve problems. That is the real relevance of technology.

CHAPTER 3

TECHNOLOGY IN THE NEW ERA

Technology has been a significant part of the transition into the New Age. Thanks to the Internet and social media, Lightworkers are able to share channelled information and new ideas with those around the globe who are seeking a higher vibration. Soul families are being united and other dimensional beings are able to share their messages with those meant to hear them. To understand the power of what is happening digitally, it is essential to understand the process of creation. Even if a thought is only known to the thinker, it forever becomes part the consciousness of civilization. Since thought is energy, which cannot be destroyed, the only thing that can happen to a thought is for it to be transmuted into another form. This is accomplished through intention. If the thought is written, it becomes even more powerful, yet it can, at some point, be retracted and replaced by a different thought.

As new information regarding the ascension process is written on web pages and social media sites, the vibrations that those messages carry is integrated into the Light Quotient of humanity. Channeled messages that have a very high vibration, have the ability to exponentially enlighten those who are open to and who vibrate with the message. This can be seen by the increase in the number of Lightworkers moving into the New Era, which began in 2013; the foundation of which, is being constructed as the messages of Light become available, mainly in digital format.

Thanks to infra-red satellite technology, remains of past civilizations that are still buried, are being detected from space. These archeological discoveries expose a heritage that validates what Lightworkers already know about from information received in channels and personal memories of other incarnations. It is becoming accepted, even in forward thinking scientific circles, that there were far more advanced civilizations on Earth than the current one. What is yet to be accepted by science, is that and many who are alive today also had other lifetimes in those cultures that are long gone.

As ruins are excavated, carbon dated and the findings are returned, it is being confirmed that advanced people thrived on Earth much earlier than traditional history books claim. As technology advances, it might be a good idea for historians, who intend to re-write history, based on the new discoveries, to leave a window of opportunity open for the possibility that this is not the end of what might be found. In light of these discoveries, the word ancient takes on new meaning. Interestingly, when one takes stock of what is found in the ruins of such cultures, there is very little recorded history to be found. How is it that these mathematically advanced societies, which built with precision that surpasses that of current modern times, lasted so long, but did not amass vast libraries containing the wisdom and knowledge that had been accumulated over the centuries? Why is there not more than the few wall carvings and hieroglyphics found on the sides of buildings.

In modern day culture, it is assumed that all that is learned must be recorded for posterity. The more knowledge that is gathered in the libraries of each culture, the wiser the society considers itself. However, writing and reading are not the only forms of recording information. Ancient civilizations, truly ancient civilizations, knew of other means of communicating and preserving knowledge. Crystals were used to store information, move objects provide free, clean energy and heal the body and mind. None of these technologies have been developed by modern science. However, at the rate that the digital age is developing, many amazing changes might be right around the corner.

Indigenous cultures share their sacred wisdom, beliefs and knowledge orally from person-to-person. Happily though, this system is returning to modern-day culture in the form of the Internet and social networking, through which information is shared from person to person. Do you get it? The old ways are returning, but in a new platform. Technology is the expression of humanity's current major achievements, the next step out of the darkness and the canvas of communication of the New Era.

People are now reconnected through social media. Boundaries of all sorts are forgotten and there is a global community being formed. Do you see the beauty of it? At the same time, many are removing themselves from direct human contact with others, which has been the primary form of communication over the centuries, yet, in doing so, other methods of communication, through voice, written word and video are taking its place. Does this take something away from human relations? Yes, but it also adds something else in its place. Is one better than the other? Moving away from this linear way of thinking of which is better, allows one to realize that there are many ways for humans to connect. In fact, emotions are just as strong between people communicating in these new ways as they are meeting face to face. Since so much emphasis is placed on physical appearance in modern culture, many find it easier to be honest and open with someone they are getting to know, when it is done with the buffer of the Internet between them. The ability to tune into a person telepathically is being awakened and consequently strengthened in people as they connect with others remotely. It is just as easy to fall in love with someone on the Internet as it is in person.

As the technological phenomenon of social media takes place, new neural pathways are being created by an entire generation of predominantly young people who are placing their focus on electronic devices as they text and connect with the knowledge of the world at their fingertips. These new activities are developing skills, both physical and mental, necessary for the future development of individuals living in the modern world. At the moment, it might seem like just texting, but the parts of the brain that are being utilized by this activity will be used for technology that has yet to be discovered. Humans are not developing new technology, for all technology already exists. They are only discovering and deploying it. One breakthrough leads to the next, until little by little, in the not so distant future, there will again be an entirely new reality in place. After all, not so long ago there was no Internet and a typewriter was the means of placing words on paper.

Until the last century, education was not common amongst the masses and information was only available to a select few, but today, thanks to technology, anyone with an Internet connection can get an education. Major universities offer free online classes and virtually any subject can be researched on the web. This has resulted in rapid evolutionary shifts in previously uneducated sectors of the population. The results can be seen in the massive protests, organized on the Internet in very short periods of time. Entire governments can be toppled by a Facebook group. It used to be that these sort of revolutions took much time, planning, preparation and years of struggle. What was once a long process can now be accomplished in days-power is being returned to the people.

Technology is changing and evolving in proportion to the rate of Conscious Human Evolution, there's no stopping it. There is no precedent that we know of upon which to measure the progress, no guidebooks and no manuals. We are in uncharted territory. Yet, those aware of the New Era, living outside the realm of the collective consciousness, are assisted by the technology that accompanies their advancement. There is definitely a plan within the creative consciousness of the universe to assist mankind in rising to heights that it has not seen for eons. At this time, there are beings whose purpose it is to assist humanity in its development and gratitude must be given to those that are sending Light and transferring information through Lightworkers who are channeling them. There is no denying that the new information is coming from an extraterrestrial source. The term extraterrestrial is meant to describe not only beings living on other planets, but all those living outside the third dimension on Earth, including Angels, Ascended Masters and other beings of light. The vibration of the messages that are received from them is an accelerator to the process of moving into the New Era.

What the technology of the future will bring, we can only imagine; however, remember this, what can be created with the mind, can also be created in the Third Dimension and as the overall vibration of humanity becomes higher, people will be able to accomplish things that are currently believed to be impossible. An indicator of what is in store can be found in movies. Fiction is an account of what is happening in another dimension. The writer is connected through a thinning of the veil to another place, tapping into the story. Once it is brought into the consciousness of millions of people, it then becomes part of reality. Even though it is considered it to be fiction, it is now added to the collective consciousness. It is no coincidence that stories like Harry Potter, Avatar, Once Upon a Time are becoming mainstream at this time. The fairy tales and legends that ever have been told are all account of events that took place in another time. They are tales of our multidimensional heritage. There really is no such thing as fiction.

It was necessary for a substantial amount of people to evolve to a higher level by 2013 so that humanity could step into its new energetic platform. This has occurred. The New Era has been created and its foundation is the electromagnetic grid of the earth, which after eons is again intact. This ensures that never again will humanity fall to the level it had in the last era. Now, finally, those who have embraced the new reality will create a system of living on Earth that will benefit all. Technology is the platform of the New Era that is being built. Embrace it, for it is your future.

EDUCATIONAL TECHNOLOGY

There is no written evidence which can tell us exactly who has coined the phrase educational technology. Different educationists, scientists and philosophers at different time intervals have put forwarded different definitions of Educational Technology. Educational technology is a multifaceted and integrated process involving people, procedure, ideas, devices, and organization, where technology from different fields of science is borrowed as per the need and requirement of education for implementing, evaluating, and managing solutions to those problems involved in all aspects of human learning.

EDUCATIONAL TECHNOLOGY, BROADLY SPEAKING, HAS PASSED THROUGH FIVE STAGES.

The first stage of educational technology is coupled with the use of aids like charts, maps, symbols, models, specimens and concrete materials. The term educational technology was used as synonyms to audio-visual aids.

The second stage of educational technology is associated with the 'electronic revolution' with the introduction and establishment of sophisticated hardware and software. Use of various audio-visual aids like projector, magic lanterns, tape-recorder, radio and television brought a revolutionary change in the educational scenario. Accordingly, educational technology concept was taken in terms of these sophisticated instruments and equipments for effective presentation of instructional materials.

The third stage of educational technology is linked with the development of mass media which in turn led to 'communication revolution' for instructional purposes.

Computer-assisted Instruction (CAI) used for education since 1950s also became popular during this era.

The fourth stage of educational technology is discernible by the individualized process of instruction. The invention of programmed learning and programmed instruction provided a new dimension to educational technology. A system of self-learning based on self-instructional materials and teaching machines emerged.

The latest concept of educational technology is influenced by the concept of system engineering or system approach which focuses on language laboratories, teaching machines, programmed instruction, multimedia technologies and the use of the computer in instruction. According to it, educational technology is a systematic way of designing, carrying out and evaluating the total process of teaching and learning in terms of specific objectives based on research.

EDUCATIONAL TECHNOLOGY DURING THE STONE AGE, THE BRONZE AGE, AND THE IRON AGE

Educational technology, despite the uncertainty of the origin of the term, can be traced back to the time of the three-age system periodization of human prehistory; namely the Stone Age, the Bronze Age, and the Iron Age.

During the Stone Age, ignition of fire by rubbing stones, manufacture of various handmade weapon and utensils from stones and clothing practice were some of the simple technological developments of utmost importance. A fraction of Stone Age people developed ocean-worthy outrigger canoe ship technology to migrate from one place to another across the Ocean, by which they developed their first informal education of knowledge of the ocean currents, weather conditions, sailing practice, astronavigation, and star maps. During the later Stone Age period (Neolithic period),for agricultural practice, polished stone tools were made from a variety of hard rocks largely by digging underground tunnels, which can be considered as the first steps in mining technology. The polished axes were so effective that even after appearance of bronze and iron; people used it for clearing forest and the establishment of crop farming.

Although Stone Age cultures left no written records, but archaeological evidences proved their shift from nomadic life to agricultural settlement. Ancient tools conserved in different museums, cave paintings like Altamira Cave in Spain, and other prehistoric art, such as the Venus of Willendorf, Mother Goddess from Laussel, France etc. are some of the evidences in favour of their cultures.

Neolithic Revolution of Stone Age resulted into the appearance of Bronze Age with development of agriculture, animal domestication, and the adoption of permanent settlements. For these practices Bronze Age people further developed metal smelting, with copper and later bronze, an alloy of tin and copper, being the materials of their choice.

The Iron Age people replaced bronze and developed the knowledge of iron smelting technology to lower the cost of living since iron utensils were stronger and cheaper than bronze e☐uivalents. In many Eurasian cultures, the Iron Age was the last period before the development of written scripts.

EDUCATIONAL TECHNOLOGY DURING THE PERIOD OF ANCIENT CIVILIZATIONS

According to Paul Saettler, 2004, Educational technology can be traced back to the time when tribal priests systematized bodies of knowledge and ancient cultures invented pictographs or sign writing to record and transmit information. In every stage of human civilization, one can find an instructional technique or set of procedures intended to implement a particular culture which were also supported by number of investigations and evidences. The more advanced the culture, the more complex became the technology of instruction designed to reflect particular ways of individual and social behaviour intended to run an educated society. Over centuries, each significant shift in educational values, goals or objectives led to diverse technologies of instruction.

The greatest advances in technology and engineering came with the rise of the ancient civilizations. These advances stimulated and educated other societies in the world to adopt new ways of living and governance.

The Indus Valley Civilization was an early Bronze Age civilization which was located in the northwestern region of the Indian Subcontinent. The civilization was primarily flourished around the Indus River basin of the Indus and the Punjab region, extending upto the Ghaggar-Hakra River valley and the Ganges-Yamuna Doab, (most of the part is under today's Pakistan and the western states of modern-day India as well as some part of the civilization extending upto southeastern Afghanistan, and the easternmost part of Balochistan, Iran).

There is a long term controversy to be sure about the language that the Harappan people spoke. It is assumed that their writing was at least seems to be or a pictographic script. The script appears to have had about 400 basic signs, with lots of variations. People write their script with the direction generally from right to left. Most of the writing was found on seals and sealings which were probably used in trade and official & administrative work.

Harappan people had the knowledge of the measuring tools of length, mass, and time. They were the first in the world to develop a system of uniform weights and measures.

In a study carried out by P. N. Rao et al. in 2009, published in Science, computer scientists found that the Indus script's pattern is closer to that of spoken words, which supported the proposed hypothesis that it codes for an as-yet-unknown language.

According to the Chinese Civilization, some of the major techno-offerings from China include paper, early seismological detectors, toilet paper, matches, iron plough, the multi-tube seed drill, the suspension bridge, the wheelbarrow, the parachute, natural gas as fuel, the magnetic compass, the raised-relief map, the blast furnace, the propeller, the crossbow, the South Pointing Chariot, and gun powder. With the invent of paper they have given their first step towards developments of educational technology by further culturing different handmade products of paper as means of visual aids.

Ancient Egyptian language was at one point one of the longest surviving and used languages in the world. Their script was made up of pictures of the real things like birds, animals, different tools, etc. These pictures are popularly called hieroglyph. Their language was made up of above 500 hieroglyphs which are known as hieroglyphics. On the stone monuments or tombs which were discovered and rescued latter on provides the evidence of existence of many forms of artistic hieroglyphics in ancient Egypt.

EDUCATIONAL TECHNOLOGY DURING MEDIEVAL AND MODERN PERIOD

Paper and the pulp papermaking process which was developed in China during the early 2nd century AD, was carried to the Middle East and was spread to Mediterranean by the Muslim conquests. Evidences support that a paper mill was also established in Sicily in the 12th century. The discovery of spinning wheel increased the productivity of thread making process to a great extent and when Lynn White added the spinning wheel with increasing supply of rags, this led to the production of cheap paper, which was a prime factor in the development of printing technology.

The invention of the printing press was taken place in approximately 1450 AD, by Johannes Gutenburg, a German inventor. The invention of printing press was a prime developmental factor in the history of educational technology to convey the instruction as per the need of the complex and advanced-technology cultured society.

In the pre-industrial phases, while industry was simply the handwork at artisan level, the instructional processes were relied heavily upon simple things like the slate, the horn book, the blackboard, and chalk. It was limited to a single text book with a few illustrations. Educational technology was considered synonymous to simple aids like charts and pictures.

The year 1873 may be considered a landmark in the early history of technology of education or audio-visual education. An exhibition was held in Vienna at international level in which an American school won the admiration of the educators for the exhibition of maps, charts, textbooks and other equipments.

Maria Montessori (1870-1952), internationally renowned child educator and the originator of Montessori Method exerted a dynamic impact on educational technology through her development of graded materials designed to provide for the proper sequencing of subject matter for each individual learner. Modern educational technology suggests many extension of Montessori's idea of prepared child centered environment.

In1833, Charles Babbage's design of a general purpose computing device laid the foundation of the modern computer and in 1943, the first computing machine as per hi design was constructed by International Business Machines Corporation in USA. The Computer Assisted instruction (CAI) in which the computer functions essentially as a tutor as well as the Talking Type writer was developed by O.K. Moore in 1966. Since 1974, computers are interestingly used in education in schools, colleges and universities.

In the beginning of the 19th century, there were noteworthy changes in the field of education. British Broadcasting Corporation (BBC), right from its start of school broadcasts in 1920 had maintained rapid pace in making sound contribution to formal education. In the USA, by 1952, 20 states had the provision for educational broadcasting. Parallel to this time about 98% of the schools in United Kingdom were equipped with radios and there were regular daily programmes.

Sidney L. Pressey, a psychologist of Ohio state university developed a self-teaching machine called 'Drum Tutor' in 1920. Professor Skinner, however, in his famous article 'Science of Learning and art of Teaching' published in 1945 pleaded for the application of the knowledge derived from behavioral psychology to classroom procedures and suggested automated teaching devices as means of doing so.

Although the first practical use of Regular television broadcasts was in Germany in 1929 and in 1936 the Olympic Games in Berlin were broadcasted through television stations in Berlin, Open circuit television began to be used primarily for broadcasting programmes for entertainment in 1950. Since 1960, television is used for educational purposes.

In 1950, Brynmor, in England, used educational technological steps for the first time. It is to be cared that in 1960, as a result of industrial revolution in America and Russia, other countries also started progressing in the filed of educational technology. In this way, the beginning of educational technology took place in 1960 from America and Russia and now it has reached England, Europe and India.

During the time of around 1950s, new technocracy was turning it attraction to educations when there was a steep shortage of teachers in America and therefore an urgent need of educational technology was felt. Dr. Alvin C. Eurich and a little later his associate, Dr. Alexander J. Stoddard introduced mass production technology in America.

Team teaching had its origin in America in the mid of 1950's and was first started in the year 1955 at Harvard University as a part of internship plan.

In the year 1956, Benjamin Bloom from USA introduced the taxonomy of educational objectives through his publication, "The Taxonomy of Educational Objectives, The Classification of Educational Goals, Handbook I: Cognitive Domain".

In 1961, Micro teaching technique was first adopted by Dwight W. Allen and his co-workers at Stanford University in USA.

Electronics is the main technology being developed in the beginning of 21st century. Broadband Internet access became popular and occupied almost all the important offices and educational places and even in common places in developed countries with the advantage of connecting home computers with music libraries and mobile phones.

Today's classroom is more likely to be a technology lab, a room with rows of students using internet connected or Wi-Fi enabled laptops, palmtops, notepad, or perhaps students are attending a video conferencing or virtual classroom or may have been listening to a podcast or taking in a video lecture. Rapid technological changes in the field of educational have created new ways to teach and to learn. Technological changes also motivated the teachers to access a variety of information on a global scale via the Internet, to enhance their lessons as well as to make them competent professional in their area of concern. At the same time, students can utilize vast resources of the Internet to enrich their learning experience to cope up with changing trend of the society. Now a days students as well teachers are attending seminars, conferences, workshops at national and international level by using the multimedia techno-resources like PowerPoint and even they pursue a variety of important courses of their choice in distance mode via online learning ways. Online learning facility has opened infinite number of doors of opportunities for today's learner to make their life happier than ever before.

LEAPING INTO THE 6TH TECHNOLOGY REVOLUTION

We're at risk of missing out on some of the most profound opportunities offered by the technology revolution that has just begun.

Yet many are oblivious to the signs and are in danger of watching this become a period of noisy turmoil rather than the full-blown insurrection needed to launch us into a green economy. What we require is not a new spinning wheel, but fabrics woven with nanofibers that generate solar power. To make that happen, we need a radically reformulated way of understanding markets, technology, financing, and the role of government in accelerating change. But will we understand the opportunities before they disappear?

SEEING THE SIXTH REVOLUTION FOR WHAT IT IS

We are seven years into the beginning of what analysts at BofA Merrill Lynch Global Research call the Sixth Revolution. A table by Carlotta Perez, which was presented during a recent BofA Merrill Lynch Global Research luncheon hosted by Robert Preston and Steven Milunovich, outlines the revolutions that are unexpected in their own time that lead to the one in which we find ourselves.

1771: Mechanization and improved water wheels

1829: Development of steam for industry and railways

1875: Cheap steel, availability of electricity, and the use of city gas

1908: Inexpensive oil, mass-produced internal combustion engine vehicles, and universal electricity

1971: Expansion of information and tele-communications

2003: Cleantech and biotech

The Vantage of Hindsight

Looking back at 1971, we know that Intel's introduction of the microprocessor marked the beginning of a new era. But in that year, this meant little to people watching Mary Tyler Moore and The Partridge Family, or listening to Tony Orlando & Dawn and Janis Joplin. People would remember humanity's first steps on the Moon, opening relations between US and China, perhaps the successful completion of the Human Genome Project to 99.99% accuracy, and possibly the birth of Prometea, the first horse cloned by Italian scientists.

According to Ben Weinberg, Partner, Element Partners, "Every day, we see American companies with promising technologies that are unable to deploy their products because of a lack of debt financing. By filling this gap, the government will ignite the mass deployment of innovative technologies, allowing technologies ranging from industrial waste heat to pole-mounted solar PV to prove their economics and gain credibility in the debt markets."

Flying beneath our collective radar was the first floppy disk drive by IBM, the world's first e-mail sent by Ray Tomlinson, the launch of the first laser printer by Xerox

PARC and the Cream Soda Computer by Bill Fernandez and Steve Wozniak (who would found the Apple Computer company with Steve Jobs a few years later).

Times have not changed that much. It's 2011 and many of us face a similar disconnect with the events occurring around us. We are at the equivalent of 1986, a year on the cusp of the personal computer and the Internet fundamentally changing our world. 1986 was also the year that marked the beginning of a major financial shift into new markets. Venture Capital (VC) experienced its most substantial finance-raising season, with approximately $750 million, and the NASDAQ was established to help create a market for these companies.

Leading this charge was Kleiner Perkins Caulfield & Beyers (KPCB), a firm that turned technical expertise into possibly the most successful IT venture capital firm in Silicon Valley. The IT model looked for a percentage of big successes to offset losses: an investment like the $8 million in Cerent, which was sold to Cisco Systems for $6.9 billion, could make up for a lot of great ideas that didn't quite make it.

CHANGING FINANCIAL MODELS

But the VC model that worked so well for information and telecommunications doesn't work in the new revolution. Not only is the financing scale of the cleantech revolution orders of magnitude larger than the last, this early in the game even analysts are struggling to see the future.

Steven Milunovich, who hosted the BofA Merrill Lynch Global Research lunch, remarked that each revolution has an innovation phase which may last for as long as 25 years, followed by an implementation phase of another 25. Most money is made in the first 20 years, so real players want to get in early. But the question is: Get in where, for how much and with whom?

There is still market scepticism and uncertainty about the staying power of the clean energy revolution. Milunovich estimates that many institutional investors don't believe in global warming, and adopt a "wait and see" attitude complicated by government impasse on energy security legislation. For those who are looking at these markets, their motivation ranges from concerns about oil scarcity, supremacy in the "new Sputnik" race, the shoring up of homeland security and - for some - a concern about the effects of climate change. Many look askance at those who see that we are in the midst of a fundamental change in how we produce and use energy. Milunovich, for all these reasons, is "cautious in the short term, bullish on the long."

THE VALLEY OF DEATH

Every new technology brings with it needs for new financing. In the sixth revolution, with budget needs 10 times those of IT, the challenge is moving from idea to prototype to commercialization. The Valley of Death, as a recent Bloomberg New Energy Finance whitepaper, Crossing the Valley of Death pointed out, is the gap between technology creation and commercial maturity.

But some investors and policy makers continue to hope that private capital will fuel this gap, much as it did the last. They express concern over the debt from government programs like the stimulus funds (American Recovery and Reinvestment Act) which have invested millions in new technologies in the clean energy sector, as well as helping states with rebuilding infrastructure and other projects. They question why the traditional financing models, which made the United States the world leader in information technology and telecommunications, can't be made to work today, if the Government would just get out of the way.

But analysts from many sides of financing believe that government support, of some kind, is essential to move projects forward, because cleantech and biotech projects require a much larger input of capital in order to get to commercialization. This gap not only affects commercialization, but is also affecting investments in new technologies, because financial interests are concerned that their investment might not see fruition - get to commercial scale.

How new technologies are radically different from the computer revolution.

Infrastructure complexity

This revolution is highly dependent on an existing - but aging - energy infrastructure. Almost 40 years after the start of the telecommunications revolution, we are still struggling with a communications infrastructure that is fragmented, redundant, and inefficient. Integrating new sources of energy, and making better use of what we have, is an even more complex - and more vital - task.

According to "Crossing the Valley of Death," the Bloomberg New Energy Finance Whitepaper,

"The events of the past few years confirm that it is only with the public sector's help that the Commercialization Valley of Death can be addressed, both in the short and the long term. Only public institutions have 'public benefits' obligations and the associated mandated risk-tolerance for such classes of investments, along with the capital available to make a difference at scale. Project financiers have shown they are willing to pick up the ball and finance the third, 23rd, and 300th project that uses that new technology. It is the initial technology risk that credit committees and investment managers will not tolerate."

Everything runs on fuel and energy, from our homes to our cars to our industries, schools, and hospitals. Most of us have experienced the disconnect we feel when caught in a blackout: "The air-conditioner won't work so I guess I'll turn on a fan," only to realize we can't do either. Because energy is so vital to every aspect of our economy, federal, state and local entities regulate almost every aspect of how energy is developed, deployed, and monetized. Wind farm developers face a patchwork quilt of municipal, county, state and federal regulations in getting projects to scale.

Incentives from government sources, as well as utilities, pose both an opportunity and a threat: the market rises and falls in direct proportion to funding and incentives. Navigating these challenges takes time and legal expertise: neither of which are in abundant supply to entrepreneurs.

DEVELOPMENT COSTS

Though microchips are creating ever-smaller electronics, cleantech components - such as wind turbines and photovoltaics - are huge. They can't be developed in a garage, like Hewlett and Packard's first oscilloscope. A new generation of biofuels that utilizes nanotechnology isn't likely to take place out of a dorm room, as did Michael Dell's initial business selling customized computers. What this means for sixth revolution projects is that they have much larger funding needs, at much earlier stages.

Stepping up and supporting innovation, universities - and increasingly corporations - are partnering with early stage entrepreneurs. They are providing technology resources, such as laboratories and technical support, as well as management expertise in marketing, product development, government processes, and financing. Universities get funds from technology transfer arrangements, while corporations invest in a new technologies, expanding their product base, opening new businesses, or providing cost-benefit and risk-analysis of various approaches.

But even with such help, venture capital and other private investors are needed to augment costs that cannot be born alone. These investors look to some assurance that projects will produce revenue in order to return the original investment. So concerns over the Valley of Death affects even early stage funding.

TIME LINE TO COMPLETION

So many of us balk at two year contracts for our cell phones that there is talk of making such requirements illegal. But energy projects, by their size and complexity, look out over years, if not decades. Commercial and industrial customers look to spread their costs over ten to twenty years, and contracts cover contingencies like future business failure, the sale of properties, or the prospect of renovations that may affect the long term viability of the original project.

Kevin Walsh, managing director and head of Power and Renewable Energy at GE Energy Financial Services states, "GE Energy Financial Services supports the creation of CEDA or a similar institution because it would expand the availability of low-cost capital to the projects and companies in which we invest, and it would help expand the market for technology supplied by other GE businesses."

Michael Holman, analyst for Lux Research, noted that a $25 million investment in Google morphed into $1.7 billion 5 years later. In contrast, a leading energy storage company started with a $300 million investment, and 9 years later valuation remains uncertain. These are the kinds of barriers that can stall the drive we need for 21st century technologies.

Looking to help bridge the gap in new cleantech and biotech projects, is a proposed government-based solution called the Clean Energy Deployment Administration (CEDA). There is a house and senate version, as well as a house Green Bank bill to provide gap financing. Recently, over 42 companies, representing many industries and organizations, signed a letter to President Obama, supporting the Senate version, the "21st Century Energy Technology Deployment Act."

Both the house and senate bills propose to create, as an office within the US Department of Energy (DOE), an administration which would be tasked with lending to risky cleantech projects for the purpose of bringing new technologies to market. CEDA would be the bridge needed to ensure the successful establishment of the green economy, by partnering with private investment to bring the funding needed to get these technologies to scale. Both versions capitalize the agency with $10 Billion (Senate) and $7.5 Billion (House), with an expected 10% loss reserve long term.

By helping a new technology move more effectively through the pipeline from idea to deployment, CEDA can substantially increase private sector investment in energy technology development and deployment. It can create a more successful US clean energy industry, with all the attendant economic and job creation benefits.

WHO BENEFITS?

CEDA funding could be seen as beneficial for even the most unlikely corporations. Ted Horan is the Marketing and Business Development Manager for Hycrete, a company that sells a waterproof concrete. Hardly a company that springs to mind when we think about clean technologies, he recently commented on why Hycrete CEO, Richard Guinn, is a signatory on the letter to Obama:

"The allocation of funding for emerging clean energy technologies through CEDA is an important step in solving our energy and climate challenges. Companies on the cusp of large-scale commercial deployment will benefit greatly and help accelerate the adoption of clean energy practices throughout our economy."

In his opinion, the manufacturing and construction that is needed to push us out of a stagnating economy will be supported by innovation coming from the cleantech and biotech sectors.

Google's Dan Reicher, Director of Climate Change and Energy Initiatives, has been a supporter from the inception of CEDA. He has testified before both houses of Congress, and was a signatory on the letter to President Obama. Google's interest in clean and renewable energies dates back several years. The company is actively involved in projects to cut costs of solar thermal and expand the use of plug-in vehicles, and has developed the Power Meter, a product which brings home energy management to anyone's desktop- for free.

Financial support includes corporations like GE Energy Financial Services, Silicon Valley Venture Capital such as Kleiner, Perkins Caulfiled and Byers, and Mohr Davidow Ventures, and Energy Capital including Hudson Clean Energy and Element Partners.Can something like the senate version of CEDA leap the Valley of Death?

As Will Coleman from Mohr Davidow Ventures, said, "The Devil's in the details." The Senate version has two significant changes from previous proposals: an emphasis on breakthrough as opposed to conventional technologies, and political independence.

Neil Auerbach, Managing Partner, Hudson Clean Energy

The clean energy sector can be a dynamic growth engine for the US economy, but not without thoughtful government support for private capital formation. **[Government policy] promises to serve as a valuable bridging tool to accelerate private capital formation around companies facing the challenge, and can help ensure that the US remains at the forefront of the race for dominance in new energy technologies.

BREAKTHROUGH TECHNOLOGIES

Coleman said that "breakthrough" includes the first or second deployment of a new approach, not just the game changing science-fiction solution that finally brings us limitless energy at no cost. The Bloomberg New Energy white paper uses the term "First of Class." Bringing solar efficiency up from 10% to 20%, or bringing manufacturing costs down by 50%, would be a breakthrough that would help us begin to compete with threats from China and India. Conventional technologies, those that are competing with existing commercialized projects, would get less emphasis.

POLITICAL INDEPENDENCE

Political independence is top of mind for many who spoke or provided an analysis of the bill. Michael Holman, analyst at Lux Research, expressed the strongest concerns that CEDA doesn't focus enough on incentives to bring together innovative start-ups with larger established firms.

"The government itself taking on the responsibility of deciding what technologies to back isn't likely to work- it's an approach with a dreadful track record. That said, it is important for the federal government to lead - the current financing model for bringing new energy technologies to market is broken, and new approaches are badly needed."

For many, the senate bill has many advantages over the house bill, in providing for a decision making process that includes technologists and private sector experts.

"I think both sides [of the aisle] understand this is an important program, and must enable the government to be flexible and employ a number of different approaches. The Senate version empowers CEDA to take a portfolio approach and manage risk over time, which I think is good. In the House bill, CEDA has to undergo the annual appropriation process, which runs the risk of politicizing every investment decision in isolation and before we have a chance to see the portfolio mature." - Will Coleman, Mohr Davidow.

Michael DeRosa, Managing Director of Element Partners added,

"The framework must ensure the selection of practical technologies, optimization of risk/return for taxpayer dollars, and appropriate oversight for project selection and spending. **Above all, these policies must be designed with free markets principles in mind and not be subject to political process."

If history is any indication, rarely are those in the middle of game-changing events aware of their role in what will one day be well-known for their sweeping influence. But what we can see clearly now is the gap between idea and commercial maturity. CEDA certainly offers some hope that we may yet see the cleantech age grow up into adulthood. But will we act quickly enough before all of the momentum and hard work that has brought us this far falls flat as other countries take leadership roles, leaving us in the dust?

CHAPTER 4

THE SOCIAL TECHNOLOGICAL LANDSCAPE

A DISCOURSE ON TECHNO-SOCIOLOGICAL BEHAVIORS

Technological advances are no longer terms that prompt confused facial responses and infantile explanations. The presence of these advances and a host of recreational gadgets transform mediocrity into fame. These projections are witnessed through television documentaries, motion pictures, and supportive media. The issues raised by technological advances guide the process of social gentrification. This is revealed by the attention paid to the definition of "technology". There is variety of newly constructed social settings juxtaposed to an already delineated environment. This produces and articulates an enticing arrangement of social interaction. The appearance of acceptability and stability presented by these various mediums is void of theoretical development as a course of change agency. The social fabric of human interaction is achieved by re materializing loyalties of a new milieu toward immaterial cultural practices and fixation on the politics of identity. The immense influence of this societal and cultural movement towards technology, substitutes our attention from social principals and relations to behavior that taints social acceptance. The number of individuals who appear to be outwardly secure in the world of cell phones, game boys, and iPods is emergent. This conduct has displaced human elements of decision making based on firsthand experience and social contacts that would naturally materialize. Modern youth are not involved in social activities benefiting from the differences that various cultures have afforded. Ones proclivities are such that these isolated and collective social contacts produce the understanding possible to make connections that transcend the idiomatic behaviors of classes, consequential relationships, and acquaintances. Within this framework the complex manner assists in expounding everyday social life through the embodiment of meanings, values, and symbolism.

The internet enjoys marked advances adding to the extent of global reach with worldwide web and wireless communications. Themes of debate emerge citing concerns of privacy, commerce, and security as an irreversible effect on the landscape of business and personal communication, as empirical proof to the state 21st century privacy is a direct result of our technological advances. When examining the impact of technology on the application of old laws and new technologies we find that there is a 'wild west" style of social networking such as Facebook and MySpace yielding differentiated values and colloquial identity amalgamated under the technological umbrella. Considering the liberal humanism in which our young and their colleagues are engaged, the peopling of gadgets amidst a human landscape has lead to a more insensitive incorporation of technology and human agency. Very few of the individuals engaged in the peopling of technology actually represent their theory in practice.

The media, which serve as another stimulant in the lives of today's youth and adult information, are in part responsible for communicating, "poor communication." While lower standards are set by cutting edge media stars spoon feeding the illicit while sustaining these representations in the context of technology alluding to its affects as a social difference. Technological disadvantages and the isolation it creates, is a global phenomenon with local expressions. As the technological disadvantaged are labeled as unskilled labor, rural and urban America attempt to reveal the impetus behind the behaviors of anti social fears presented by its backlash

Within the clinical settings, beauty parlors, corner stores, and restaurants individuals share information and tell stories conveying verbal preservation of folklore with patrons engrossed in purveying stories to substantiate declamatory accolades. We find ourselves in a new form of discrimination through technology. This affection for privacy may not be exclusive, but the desire to escape the routine occurs without creating new ways to affix it in a subjective manner. For some the rejection of technology is a rejection of western values.

The individualism and lack of communal effort can be realized within both, individuals who arrive from the suburbs, and the ever present factions in the city who feel you should be privy to their young families, foul language, and dirty laundry. The separable variables, iPods, make it easy to be individually and silently plugged into various modes of pacification while trenched in dominant ideals of suburban life, and the expression for new conditions of experience by a consolidation of new technological socialization initiated by an emerging influence of those who turn toward gangs, and those who adhere to a resurgence of separatist behaviors.

We would be coaxed into believing this a natural order of progression in the human condition to exist positively affected by technological toys. Individuals exercise their right in taking back their privacy and peace of mind through escapism. The "let's not be here now" approach to problems posed in the urban environment is dealt with through the personal head set. In the populist language of the future, "get away from it all" are attempts to hide the erosion of family values and neighborhood security. Technology has acquired many American jobs ushered to overseas as off shoring accounts simplified through use of technology fueling revanchist behavior by the ever increasing manifestation of cheap labor. An entity technology can bring to your door step.

Amidst another technological renaissance, consequent social behavior justifies a critique of the development and political contributions of the largest market comprising our private, leisure, and employment milieu. The dissociative and apathetic behavior of a transparent human social contract is sensitive to circumstances that promote the distancing of our next door neighbor, friends, and colleagues through technological mediums. Significant growth of the technological phenomenon since its inception is illustrated by the creators of Buck Rodgers and Dick Tracy to public agencies who tend to the aftermath these two forces generate. The difficulty of procuring a format to study technological continentalism, its cause and effect, and potential to impact the behavior of society through technology is a difficult one.

HOW TECHNOLOGY CREATES WEALTH

Dynamic markets create opportunity

Markets create energy because they are dynamic. They are constantly evolving in response to changes in the economic, political and technological environments. Understanding what causes a market to evolve helps you predict where opportunities will emerge; how fast they will develop, and when and whether mass adoption will occur. If you can capture this energy, you can use it to drive the sales process.

Dynamic systems create energy. If left unchecked, any systemic change tends to grow. A snowball rolling downhill gets bigger. Growth creates momentum. As the snowball grows bigger, it goes faster. Momentum creates energy. The faster the snowball rolls; the bigger it gets; the harder it hits the tree. Energy drives change. (Source The Fifth Discipline)

You can use the energy sources created by an evolving market to motivate prospects to buy your solution. Persuading people to try out a new technology is an uphill battle. You have to invest a lot of your precious energy - sales resources, capital, technical expertise, etc. - into convincing prospects they can benefit from using your technology to support their business. However, if you understand what is driving market change- an increasingly mobile workforce, higher need for personal security, faster access to global markets -

then you use the energy created by the market to motivate prospects to buy. Thus, you need to invest less of your own resources and you can sell more productively and efficiently.

TECHNOLOGY MARKETS CREATE ABUNDANCE.

There are two laws that explain why technology-enabled markets generate extraordinary amounts of energy.

1. Moore's Law predicts that technology is going to improve in the future and cost less.

2 Metcalf's Law states that technologies become more useful as more people use them.

The combination of these two laws creates an economy of abundance that is unique to technology markets. As Moore's Law predicts an endless supply of ever-increasing resources and Metcalf's Law promises that innovations will be quickly adopted, the nature of the economy changes.

Gordon Moore, the founder of Intel, said, "Every 18 months processing power doubles while the cost holds constant." The implications of Moore's Law are that every 18 months technology is going to cost half as much and be twice as powerful. Moore's Law has held true for over 30 years. Previous economies were based on the laws of scarcity, where you have a limited amount of resources and value is based on how scarce they are - gold, oil, land, etc. The more you use up the resources the less energy you have.

A technology-based economy is based on the laws of abundance. According to Moore's law, there will always be cheaper resources tomorrow. This ever-increasing pool of resources enables customers to implement new business strategies. If it isn't possible today, it will be possible tomorrow. Improved technology is constantly fueling the market, creating energy.

Furthermore, thanks to this simple formula technological obsolescence is only a few months away. Customers can never afford to sit still for fear that a competitor will be able to leapfrog ahead of them if they adopt the next generation of technology faster. This anxiety is another powerful source of energy that you can use to drive your sales.

Metcalf's Law also has a powerful effect on developing markets. Robert Metcalf, the founder of 3Com, said "New technologies are valuable only if many people use them... the utility of a network equates the square of the number of users. " This means that the more people use a technology, the more useful it becomes. If there was only one fax machine in the world, it wouldn't be useful. With two fax machines you can send mail back and forth faster and cheaper than if you send it through the post office. With 2,000,000 fax machines, you never have to wait in line at the post office again.

According to Metcalf a technology's usefulness equals the number of users squared. If two people use a fax it is four times easier than using the postal system. If 20 people use the fax machine, it is 400 times easier. This creates a geometric increase in the technology's utility, which is just another way of saying why customers would want to buy it. So if 2 people want to buy a fax machine today; 4 people will want to buy it tomorrow; 16 people will want to buy it the day after tomorrow; 256 people will want to buy it next week, and 2,147,483,648 will want to buy it by the end of the month. That is a lot of potential customers lining up to buy your product, which is what market energy is all about.

Abundance creates demand for your technology. Since technology markets create abundance they are not subject to the constraints of scarcity. They have unlimited growth potential and consequently unlimited potential to create wealth.

Janice Lawrence has advised leading edge technology companies for the past two decades on how to sell innovative technology.

CHAPTER 5

TECHNOLOGY - MEANING AND EFFECTS ON LIFE

Technology was coined from the Greek word technologia, which refers to a "skill" or an "art". It simply refers to the utilization and cognition of techniques, tools or methods of organization. It could also mean the in-depth analysis of a concept or the proper knowledge of a field or subject area. The use of the word is relative, and can be applied to as many areas as possible. Mobile technology, car technology, medical technology, space technology and a host of other terms have been coined from the word. This is because the application of the word is limitless in the present world.

The impact of all the different technologies on humans is very significant. Technology started with the transformation of natural resources into elementary hand tools. Fire, for instance, played a great role in the advancement of prehistoric technology.

Its discovery and control was the beginning of the many and has escalated to inexplicable heights like the internet today. It is pertinent to note that not all advancements in the field have fostered peace and made life easier. Nuclear weapons and their likes are of course, utterly destructive.

All the great economies in the world today depend on technology. It is safe to assume that they would not get this far but for technological advancements. There is also the issue of pollution and the depletion of natural resources as a result of processes that utilize technology. This reduces the ▢uality of the earth and the value of the environment in general. Interestingly, recent studies have discovered that other primates carry out activities that point to an understanding of basic technology.

Lots of debates have arisen over the years and they have centered on the impact of technology on the society. Some movements totally condemn the concept with overwhelming evidence and conviction. Different scholars and philosophers read different meanings into it and extend its definition to cover a broad range of usage. This has made science, technology and engineering overlap in their meanings and applications. The history of technology also makes for interesting reading as it dates back to 10,000 BC. It has evolved from the usage of basic tools to the development of all the crucial sectors in the life of man. Medicine, agriculture, manufacturing, transportation, communication and education, to mention a few, have benefited immensely from this mind-boggling concept.

The view on technology cannot be uniform. It always differs from scholar to scholar and from individual to individual. Where uniformity lies however, is in the fact that technology has come down a long, winding way, and that it is here to stay.

THE TECHNOLOGY OF NATIONS

In 1776, Scottish economist and philosopher, Adam Smith wrote the masterpiece, 'The Wealth of Nations'- actually 'An Inquiry into the Nature and Causes of the Wealth of Nations". By coincidence, the United States Declaration of Independence was adopted the same year, making the American colonies independent and thus no longer a part of the British Empire.

America has since evolved to dominate the old British Empire in virtually every aspect of human endeavors, except perhaps, social welfare. The Yankees figuratively were discipled by Dr. Smith who believed in free market and made his argument that 'capitalism' will benefit mankind than any other economic structure. He laid this foundation at the onset of industrial revolution and provided the basics for modern economics.

Smith made his case about the 'invisible hand' and why monopoly and undue and unfettered government regulations or interference in market and industry must be discouraged. He was of the opinion that prudent allocation of resources cannot happen when states dominate and over interfere.

In that old time, America farmers could grow cotton, but would not process it. It has to be sent to England where it would later be imported into U.S as a finished product. Understanding that this decision was not due to lack of processing ability, you will appreciate Smith's argument that market must be free.

His theses were clear and were very influential; they provided the same level of fulcrum to Economics as Isaac Newton's Mathematica Prinicipia to Physics. Or in modern times, Bill Gates' Windows to the information economy.

While reading Smith's book and understanding the time frame it was written, one cannot but appreciate the intellectual rigor in that piece. Before technology was penetrated in en mass across the regions of the world, he noted that all nations could compete at par in agricultural productivity. The reason was absence of division of labor in any subsistence farming system in the world. A farmer does everything in the farm and is not an expert in most.

Discounting fertile land, rain and other factors that could help farmers, all the farmers, from Africa to plantations in Alabama, the level of productivity was similar. Why? No specialization was employed in farming business at the time.

Fast track forward when the industrial revolution set forth. The British Empire became an engine of wealth creation through automation. It was a ▯uintessential period of unrivalled human productivity which resulted to enormous wealth created in the empire. Technology not only helped speed process execution, it helped in division of labor.

Interestingly, Dr Smith had noted that except agriculture where productivity was flat because of lack of division of labor, other industries were doing just fine. And in those industries, there were organized structures which enabled division of labor. For instance in the construction industry, there were bricklayers, carpenters, painters, and so on; but a farmer was a farmer.

As you read through Wealth of Nations and observe the 21st century, it becomes evident that technology was so influential in the last few centuries. It has changed our structures and created a new business adaptation rules like outsourcing which is indeed a new breed of division of labor.

From accumulation of stock and pricing, as explained by Dr. Smith, we see today a world where technology is shaping everything in very fundamental ways for wealth creation. In this era, it has become technology as technology translates to wealth. So, nations that focus on creating, diffusing and penetrating technology will do well.

Why? It is about national technology DNA. The more passionate and innovative nations are triumphing at the global business scene. Give me Japan and I will give you electronics. Talk about United States, I will share biotechnology and pharmaceutical technologies, and indeed every major technology. Give me China, and I will give you green technologies.

So, as nations continue to compete on the technology paradigm, we see at the highest level of success measurement an embodiment captured by technology capability. When nations are understood from the lens of their Technology Readiness Index, Knowledge Economic Index, we see that countries have become technology competing nodes. In some really poor countries with no (effectual) technology, they do not have a node and are unplugged in the sphere of global wealth creation.

Simply, it will be difficult to separate the health of any modern economy from its technology. It goes beyond the wealth of that nation to its survivability. The most advanced nations are the technology juggernauts while the least developing economics barely record any technology penetration impact. For the latter, it is like still living in the pre-industrial age Dr. Smith discussed on agriculture and division of labor where processes were inefficient.

Perhaps, this explains the efficiency in developed world in both the public and private arenas. The more technologies they diffuse, the more productive they become. In other words, show me the technology and I will tell you where the nation stands in the league of countries. Interestingly, the invention of steam engine changed the world and powered the industrial revolution. The invention of transistor transformed the 20th century and is fuelling the new innovation century.

It seems that major scientific breakthroughs bring major great countries. Let me emphasize here that some old kingdoms that ruled the world such as the old Babylon, Roman Empire, and Pharaoh's Egypt; there have been associated knowledge base that put them ahead. You cannot disassociate good crop production in River Nile to the mastery of Egyptians in inventing some sections of geometry for farming. Some of the old wars had been won by developing constructs that enabled efficient transportation of soldiers to battleground. There was science and nations were winning by using that knowledge.

In conclusion, the world has been living on technology and it is indeed defining our competitive space. As nations compete, it is technology that shapes the world with wealth as the major byproducts, in some cases. I make this case because some of the best technologies had been invented for non-wealth reasons (yes, directly). Examples include Internet and radar technologies which have created wealth and spurred commercial innovations but have military origins.

MODERN SCIENCE AND TECHNOLOGY AND THE CHALLENGES OF THIRD WORLD COUNTRIES

We live in a highly sophisticated world where everything is almost achievable. There would probably have been no changes between the world of today and that of three centuries ago if necessity and serendipitous discoveries had not driven men to achieve great things. Science and technology have had huge positive effects on every society. The world today has gone digital, even human thought. Our world has been reduced to a global village and is better for it.

The benefits of science and technology far outweigh every perceived shortcoming. Some of the biggest effects of technology are in the area of communication; through the internet and mobile phones. There is advancement of communication and expansions of economic commerce. Today we hear of information and communication technology (ICT). Any institution worth its name must have it in place to be really outstanding. Information technology has become boosted in today's generation; from the field of communication, business, education, and down to the entertainment industry. Through information technology, work performances are boosted with less effort and greater productivity by using various operations. Without computers or the internet, it will be difficult for people all over the world to get their questions answered. One may use the internet to locate a wealth of information with which to answer an essay question that may have been assigned at school, communicate with people, conduct transactions, access news, buy and advertise goods. The list is endless.

The advancement of Science and technology allow mass communication today so that we not only have the television, radio and newspaper, but even mobile phones which renders a multipurpose service; from long distance calls, listening to radio and music, playing games, taking pictures, recording voice and video, and browsing the internet. The benefits we obtain as a result of services from ICT have become widespread in our generation today. It improves the productive level of individuals and workers because People's knowledge of life beyond the area they lived in is now unlimited. This idea of mass communication also profoundly affects politics as leaders now have many ways they talk directly to the people. Apart from going on air to use radio or television, politicians resort to the social media for some of their political comments and campaign. Information about protests and revolutions are being circulated online, especially through social media. This has caused political upheavals and resulted in change of government in most countries today.

Furthermore, current global issues are much more accessible to the public. Communication has been brought also to the next level because one can find new ways to be able to communicate with loved ones at home.

Science and technology expand society's knowledge. Science helps humans gain increased understanding of how the world works, while technology helps scientists make these discoveries. Learning has maximized because of different media that are being developed which are all interactive and which bring learning experiences to the next level. Businesses have grown and expanded because of breakthroughs in advertising.

Modern technology has changed the way many companies produce their goods and handle their business. The idea and use of video and web conferencing, for instance, has helped companies remove geographical barriers and given them the opportunity to reach out to employees and clients through out the world. In today's economy, it has helped companies reduce the cost and inconveniences of travelling, allowing them to meet as often as they could like without having to worry about finding the budget to settle it. Modern technology helps companies reduce their carbon footprint and become green due to the fact that almost anything can be done from a computer.

There have been advances in medical care through the development of science and technology. Advances in medical technology have contributed immensely in extending the life span of people. People with disabilities or health problems are now more and more able to live closer to normal lives. This is because science contributes to developing medications to enhance health as well as technology such as mobile chairs and even electronics that monitor current body levels. Most devices used by the physically challenged people are customized and user friendly.

Science and technology increase road safety. Nowadays, law enforcement officers use Laser technology to detect when automobiles are exceeding speed limits. Technology has led to the development of modern machines such as cars and motorcycles which allow us to be mobile and travel freely and airplanes which travel at a supersonic speed.

Another machine, the air- conditioner, provides cool comfort, especially during hot weather. In offices where dress codes exist, people can afford to wear suits without being worried about the weather. It guarantees convenience even when the climate says otherwise.

Moreover, present day factories have modern facilities like machines and soft ware that facilitate production. These machines work with greater speed and perfection incomparable with human skills. These machines have enabled markets to have surplus products all over the world. For the software, they make it possible for machines to be programmed, for production to be regulated, to monitor the progress being recorded and so on.

Modern technology indeed has been great. For third world countries, however, it has been challenging, especially the area of production. Only consuming and not been able to manufacture does not favour any country when it comes to balance of trade. The most sensitive parts of technology are the theoretical or conceptual parts and technical parts. These are the backbone of technological development anywhere in the world. Without the ideas, there will not be technology. Third world counties need to go back to the basics, that is, to the primitive. There must be meeting ground for tradition and modern technological invention. Third world countries engage in import substitution strategy where they import half finished goods and complete the tail end of the production process domestically. Third world countries started wrongly. They started with climbing the ladder from the top which is very wrong and difficult. They thought that being able to purchase and operate modern technological products qualifies for advancement in science and technological development. This makes third world countries to be a dependent system because working in the factories are routine work and this inevitably links to the issue of the idea of technology transfer. They should seek for technological transfer, but the problem is that no nation is ready to transfer her hard earned technological knowledge to any other nation for some certain reasons which drive nations into competition; world politics and economic prowess. That is the struggle to lead or dominate other nations technologically, economically and politically. Be the first to invent new gadgets and latest electronics including those used in modern warfare, use other nations as market for finished goods, and to have a strong voice and be able to influence other countries. They should consider embarking on technological espionage so as to acquire the rudiments for technological development if they must liberate themselves from the shackles of technological domination.

In conclusion, it's not until third world countries begin to put embargo on the importation of certain electronics and mechanical goods that the necessity to be creative would replace the habit of consuming foreign products. Countries like Thailand, Burma, Brazil, and South Africa and so on, should be emulated. These countries experienced colonialism yet they did not allow it to overwhelm their creative prowess. Industry and determination saw them emerge as economic giants in the world today. Third world countries should emulate them by carrying out proper feasibility studies to ascertain which technology will suit their country; giving more financial boost to this area, training people to become experts; motivating and encouraging individuals who are naturally endowed and technologically inclined to display their bests of talents. These measures if strictly adhered to will go a long way to help the advancement of these countries in the area of science and technology. If these countries must achieve greatness before the next decade, they have to make conscious and unrelenting efforts. The time starts now! The more they delay, the more backward they become.

CHAPTER 6

HOW MUCH IS "INFORMATION TECHNOLOGY DEBT" HURTING YOUR BOTTOM-LINE?

Information Technology (IT) debt is basically the cost of maintenance needed to bring all applications up to date.

Shockingly, global "Information Technology (IT) debt" will reach $500 billion this year and could rise to $1 trillion by 2015!

But why should you take IT debt seriously and begin to take steps to eliminate this issue from your business?

According to Gartner, the world's leading information technology research and advisory company...

It will cost businesses world-wide 500 billion dollars to "clear the backlog of maintenance" and reach a fully supported current technology environment.

GARTNER SUMMARIZES THE PROBLEM BEST:

"The IT management team is simply never aware of the time scale of the problem. This problem, hidden from sight, is getting bigger every year and more difficult to deal with every year."

The true danger is that systems get out of date which leads to all kinds of costly software and hardware inefficiencies.

Your tech support provider can most likely do a better job at staying current with your computer and network environment.

HAVE THEM START TODAY BY DOCUMENTING THE FOLLOWING:

The number of applications in use

The number purchased

The number failed

The current and projected costs of both operating and improving their reliability

Are you using this powerful formula to control your technology?

There's a powerful formula I'll share with you in a moment that will help you adopt new technology faster in your business.

In business, technology encompasses Information Technology (IT), Phone Systems and Web Development.

These three layers of technology form the backbone of your business's technology environment. Why is technology adoption so important?

Without new technology adoption it's impossible for businesses to be competitive in this economy. A major role of technology is to help businesses scale, design systems, and automate processes.

Studies recently have shown that adopting technology keeps businesses leaner because entrepreneurs can do more with less.

There's evidence that new business start-ups are doing so with nearly half as many workers as they did a decade ago.

For example, Wall Street Journal's Angus Loten reported that today's start-ups are now being launched with an average of 4.9 employees.

Down from 7.5 in the 1990s, according to the Ewing Marion Kauffman Foundation, a Kansas City Research group.

In other words, technology allows businesses to expand quickly with less.

Researchers at Brandeirs University found that technology driven service businesses added jobs at a rate of 5.1% from 2001 to 2009; while employment overall dwindled by .5%.

These businesses save money, expand, and create jobs by adopting new technologies.

Are you adopting new technologies fast in your business?

Speed of technology adoption is critical to your business success.

Technology is changing the speed of business; now a whole industry might expand, mature, and die in months... not years.

There's one formula that illustrates this marriage between adopting technology and business success the best... and that's the "Optimal Technology Equation."

I recommend you adopt this powerful "Optimal Technology Equation" in your business:

- Maintenance + Planning + Innovation (Adoption)=
- Enhanced Technology Capabilities=
- Reduced Costs + Increased Production=
- Increased Profitability.

Of course, this is only a brief explanation of this invaluable formula. Be one step ahead of the competition.

MANAGING TECHNOLOGY WITHIN AN ORGANIZATION

"I am putting myself to the fullest possible use, which is all I think that any conscious entity can ever hope to do." - From the HAL 9000 computer, 2001: A Space Odyssey

When it comes to technology solutions for your business it is easy to get carried away with the latest-and-greatest gadgets and solutions. Everyone wants to have the latest shiny thing. In larger organizations, managing technology can become burdensome due to competing and duplicative technology requests. Left unfettered, the company technology platform can resemble a "spaghetti bowl" over time. Often is the case, new technology requests are submitted without any business case to support their investment.

I am a big proponent of having non-technology business leaders play an active role in the determination of the technology solutions utilized at an organization. While it is critical to include an IT perspective from a technical interface standpoint, having non-IT personnel drive technology solutions often lead to decisions based on the business needs of the organization. As such, any technology request would require a business plan to support the investment.

Form A Technology Committee: This is the start of your technology approval process. Create a technology committee that represents various personnel from cross-functional departments. Consider selecting an operations, marketing, accounting, technology and finance member to this team. This committee is charged with creating the process for submitting technology solution requests for the organization as well as providing the prioritization and ultimately, approval of the requests.

Develop A Submittal Process: Inherent in a well-thought through technology strategy for an organization is developing a process for the submission of ideas. Following the "garbage-in, garbage-out" mindset, developing a detailed process for submission will help weed out the "nice to haves" and focus the committee on real, tangible solutions. This process should not only include the technology solution identified, but as importantly, the business case for its justification. For approved projects in the queue, a monthly communication should be sent to the organization recapping the activity of the committee.

Focus Your Projects: A technology committee creates focus throughout the organization. While it would be great to have every new iteration of technology that gets released, that is impractical and costly. The committee can help with providing a high-level perspective on the entire enterprise since it is considering all requests. All to often, departmental requests have a tendency to be created in a silo, with only the impact on that department considered.

Need To Have Vs. Nice To Have: This is a biggie. It is easy to feel that an iPhone 3 becomes obsolete as soon as the iPhone 4 is released, but when the technology is run by the committee, the "nice to haves" usually fail due to a lack of business case. The committee allows the organization to run with an unbiased interference with respect to technology. The committee is charged with improving ROI on technology solutions and since it is comprised cross-departmentally, there should be no "pet" projects.

One Project, Big Picture: I have headed a technology committee in the past and the greatest "aha" moment for me was the amount of similar technology solutions that were being presented from different departments. Had all of these requests been accepted, the organization would have overspent IT dollars as well as created duplicative solutions to the same issues. The committee allows for its members to "rise above" the fray of the organization and view the technology requests in the big picture. The committee's goal was to ensure that any approved request was accretive to the overall company.

Create A Business Case: This is the best way to clear out the clutter. Ask employees what they need from a technology solution and the committee will be inundated with ideas. Ask them to submit in a business case (cost justification for the investment) along with their solution and ideas are significantly reduced. The business case for a technology solution not only helps in identifying whether the investment is worth it, but also forces the author to think about how this solution interfaces within the existing platform.

Post Analysis: Lastly, carefully measuring the business case proforma against the actual cost/return of the projects not only holds the submitter responsible, but also the committee. The goal with the post analysis isn't to "call people out", but rather provide an unbiased financial review of the project. Without this type of post analysis measurement to hold this team accountable, the committee eventually will serve no purpose.

CHAPTER 7

UTILIZING TECHNOLOGY TO IMPROVE PROFITS

If you really want to become more profitable and improve operations in your company, you have to shift your focus from the following limiting thoughts about technology.

If I buy the latest production software we will be in good shape

We don't do that here

We are unique, we don't have competition that use technology to help them generate profits

The plan is in my head, people will steal it off the computer

All I need is more sales to make more profits

You've got to get the right mindset by eliminating restricting thoughts, and then you'll be ready to improve people, processes and profitability.

Do you ever wonder how a company can start out with just one idea, a passion and a vision, then 10 to 20 years later have thousands of employees and millions in sales?

What did these companies do to become so successful

Are the owners smarter than you?

Do they work harder than you?

Did they have better equipment or people than you?

No. But they do use better technology tools to drive operation (the people and the process). Operations represent about 60% or 80% of all your overhead costs but they're the least understood by US businesses.

For decades, the Japanese have focused on operations that have driven innovation and a culture of continuous improvement. In the right small business owner hands, operations and technology can be a competitive weapon.

Now, ask yourself how can your small company--- with just a handful of employees and limited resources --- turn operations and technology applications into a powerful weapon to beat competition and learn to grow and thrive!

WHY INVEST IN TECHNOLOGY / WHAT ARE THE BENEFITS

The bottom line is, if you're suffering from tight cash flow, exhausted lines of credit and top-line growth, then you have weak operations and have underutilized the technology applications onsite or off-the-shelf that can help you.

First step to rapid profit improvement is to start by questioning your employees. They usually know where costly blocks and bottlenecks are hidden.

TECHNOLOGY CAN STORE EMPLOYEE SURVEY RESULTS THAT HELP YOU TO PLAN PROFITABILITY.

Employee Questionnaire(sample)

Are your interests and ambitions being challenged

Does each department in this company have measurable standard designed to increase profitability? Does each area have documentation of process flows and procedures of how it should work?

Does everyone in this company share the goal of improving the company profits? Does the CEO hold town hall meetings about 'planned profits'?

Are you regularly told when you do good work?

Do you get the help you need to do a good job?

As an employee, do you feel you can trust your direct supervisor/manager?

Are owner/managers open and honest with employees?

Does the company provide you with continual training in areas that will make you a better employee? Has it trained you on how to cut operating expenses or increase revenue to improve profits in your area?

Are your responsibilities generally explained, well planned and organized?

Is poor performance tolerated by management? i.e., worker performance, operations bottlenecks and customer relations.

The following are other ways business productivity software drives business processes more efficiently to gain optimal results:

CREATE AN OPEN AND COMMUNICATIVE ENVIRONMENT.

By storing appraisal information within a formal database, managers can more easily communicate business strategy and create measurable goals for their employees that will support overall company objectives. In allowing employees to see the whole picture and understand better how individual goals fit into the company's business objectives. This can create a energized and engaged employees, thereby raising the business productivity of the company.

MOTIVATE YOUR EMPLOYEES USING TECHNOLOGY.

Based upon the information gathered in an online performance evaluation, managers can compare current skills with those required for advancement or other recognition or reward opportunities that present themselves as the manager tracks progress on employee goals throughout the year. You may also find you need to redirect employees to different departments if you feel their business productivity could increase elsewhere. If there are impediments to better performance, the company should review why it is happening and try to eliminate these through better allocation of resources or additional training.

MONITOR BUSINESS PRODUCTIVITY AND EMPLOYEE PROGRESS ON GOALS.

Business productivity software solutions enable managers to more easily track progress during every phase of goal completion and offer immediate reinforcement or coaching to keep performance and deadlines on track in daily operations, and utilize performance measurements for strategic planning.

ELECTRONIC COMMERCE

There are many business applications related to e-commerce, from setting up your online storefront to managing your supply chain to marketing your products and services. These technologies fall into three main categories:

BUSINESS TO BUSINESS(B2B)

Purchasing indirect supplies

Look for catalogue-based websites offered by suppliers for corporate purchases, similar to business-to-customer websites, for purchasing indirect supplies such as office furniture, pens, paper, and general office equipment.

Leveraging your existing Web presence

Improve your existing business-to-customer e-commerce website. Greater sophistication can be added into your online store to target your business clientele.

Business to Customer(B2C)

The global reach of the Internet has allowed many businesses to sell their products and services online, both at home and abroad. An electronic storefront is a website with many pre-built e-commerce components like electronic shopping carts and secure payment gateways that you can use to set up an online store.

INTERNET MARKETING

Everything you do to promote your business online is Internet marketing. For example, Internet marketing strategies include (but are not limited to) website design and content, search engine optimization, directory submissions, reciprocal linking strategies, online advertising, and email marketing.

HOW TO IMPLEMENT TECHNOLOGY TO INCREASE PROFITS

IT implementation can be a valuable tool for increasing workplace productivity, but without a careful selection of the right technologies for your specific industry and comprehensive employee training, it can also serve to reduce productivity, profitability and employee satisfaction. The return on investment will depend on whether the technologies implemented are right for a given business' needs and how prepared employees are to use them.

STEP 1

Brainstorm a list of business process improvements you may be able to realize from a technological implementation. Your list should include three categories: improvements that you know to be possible, and which are core requirements for your expense; a wish list of things you would like to have, but which may be future development efforts; and a list of things which would transform the way you do business, but which may not be possible. These three targets provide you with a present-day implementation goal, as well as a future development target--and it may be that your transformational goals could be far easier to reach than you expect.

STEP 2

Determine whether you intend to develop these technologies using in-house resources, or through outside consultancies. Nearly every major workflow technology requires extensive customization, implementation procedures and training. Small businesses can sometimes get by cheaply using staff members technologically proficient--but mistakes made at the beginning of the process can ramp up costs later on when you turn to professional outside support.

STEP 3

Avoid specifying particular technologies if you do not have the technical expertise to evaluate them properly. The purpose of the managerial process at this stage is to define goals and budgetary constraints; non-technical managers who wed themselves to specific technologies too early can miss out on substantial cost savings, and choose a technology not the best suited for the work.

STEP 4

Circulate your request for proposals among outside consultants and implementors, or establish an internal process for doing the same among your staff if you are keeping the work in-house. Major technological implementations will not succeed if they are added to the existing workload of an employee. Proper technological implementations can be more than a full-time job in and of themselves. Staff members shifted to technology implementation should have their existing duties moved to other staff resources.

STEP 5

Negotiate a time frame, budget and implementation benchmarks with your external or internal staff resources. If you are working with an outside consultant, your contract should include protections against running over budget and over schedule. Likewise, the consultant will protect his own firm by setting specific terms of the work to be completed, and charging you extra if you change them over the course of the contract.

STEP 6

Develop an implementation timetable, including the following steps: test deployment to review the work; training, if necessary; a transition phase from the current workflow to the new technology; and production deployment of the completed technology. This last phase is typically followed by an iterative process, in which improvements to the technology are collected from the staff who have direct experience working with it. When budget and time allow for it, apply a new cycle of upgrades to your technology to ensure that you are getting the most out of it.

CHAPTER 8

HOW LIFE WOULD BE AFFECTED IF TECHNOLOGY WAS TAKEN AWAY

Advances in technology improve the material's properties and changes in modern lives of humanity. Wherein the olden times, animals or physical attachments are involved when technology was not yet developed and recognized. It only means technology materials are like an endless process in which the more years to come, the more technology products will come.

This is a big advantage for us this new generation and for the future's progression. That's why technology can be classified as needs and wants of every people. Needs because of everyday used and wants because of it's irresistible features. It may seems technology evolves around the globe. So, what's the big deal if technology is out of nowhere? Back to the old style operation where we are the one who actually works the stuff we used to do and no technologies help to accompany us? You may say yes but that's gonna be difficult to handle with especially those technology dependent.

It must be true that without technology our lives going to be messy and lonely. We should consider that possibility. It's not only because of our dependence but it is the reality of what the technology brought to the daily activities of humankind before and until now. Still it will change and become more advantage in our upcoming years. Then it is more useful and approachable in nature. Lastly, everything we see and used usually is a product of technology and technological products are necessary to give more convenient and comfortable life for everybody.

TECHNOLOGY IS DISRUPTIVE - AND EMPOWERING

Technology changes the way we work, live our lives, and have fun. Technology can empower businesses with improvements in productivity, faster development and production cycles, superior decision making by employees, and enhanced customer service. But deriving these benefits from incorporating new technology is not always a smooth process. Technology is often, at first, disruptive before it becomes empowering.

Although the ideas developed in this article may have general applicability, they are mainly intended to relate to the incorporation of new information and communications technologies into business processes. Information technologies involve computers and their peripheral e☐uipment as well as the data flow across local area networks. Communications involve any voice and video activity including the telephone system and related equipment as well as the communications pathways creating the wide area networks.

TECHNOLOGY CHANGES BUSINESS PROCESSES

Every action conducted within a business is part of one process or another. Sometimes the processes are easily defined and readily observable, as in the path of a purchase order. At other times, the process is not so clear but nevertheless it still exists even if by default.

NEW TECHNOLOGIES ARE INTRODUCED INTO BUSINESS TO:

Speed up existing processes

Extend the capabilities of existing processes

Change the processes

In changing the processes, the new technologies will often allow new ways of conducting business that were not previously possible.

Other than simply speeding up existing processes, new technologies will be disruptive when first introduced. This results from having to change patterns of behavior and/or relationships with others. When disruption occurs, productivity often suffers at first, until such time as the new processes become as familiar as the old ones. At this point, hopefully, the goal has been achieved of reaching a higher level of productivity than the level at which it started before the introduction of the new technology.

THEREFORE A COMMON CYCLE THAT OCCURS WITH THE INTRODUCTION OF NEW TECHNOLOGIES INCLUDES:

Disruption

Lower productivity, and, finally,

A higher plateau of productivity than the starting point

The obvious goals for introducing new technologies are to:

Minimize the disruption

Minimize the time it takes to increase productivity

Maximize the gain in productivity

In achieving these goals it is helpful to understand the:

Context in which the processes operate, that is, who will be impacted by changes in the specific processes affected

Democratizing potential of technology

Types of people that will react in very different ways to new technologies

The processes by which a company operates and the introduction of new technologies do not exist in isolation. Both of these exist within a context that may be a part of and affect:

The social relationships within an organization and possibly with companies with whom you conduct business

Political (power) structures within an organization

How individuals view themselves and their abilities

Technology can be democratizing. If it is used to create and disseminate information useful to the mission and goals of the business, it can be a great equalizer between "levels" of management and staff. The key word is "disseminate." If access to the information is decentralized, and easy communication of the information is allowed, then "front line" workers can improve the quantity and quality of decisions they make without having to involve layers of management.

TYPES OF PEOPLE FROM A TECHNOLOGY PERSPECTIVE

From a perspective of introducing new technology into your company, you may find it helpful to understand the following four types of people:

Innovators/embracers

Enthusiasts

Acceptors

Naysayers

Innovators/embracers will investigate new technologies on their own. They will sometimes be helpful to introducing new technologies that would otherwise not have been known to the company. They will sometimes be a "thorn" in pushing for new technologies they think will be useful (or just "neat" to have) but do not fit the company's agenda or objectives. These people will embrace new technologies when introduced by others, will often be the first ones to fully incorporate and make use of it, and could help others to fully utilize new technologies.

Enthusiasts will accept new technology enthusiastically. They won't usually seek it out but will be eager to incorporate it into their processes where appropriate. As a result of their openness, they will often readily learn how to use the new technology and may also be useful in assisting others through the learning process.

Acceptors will accept new technology because it is required. They will not seek it out. In fact, they will often try to avoid it at first until they are forced to accept it. Once they understand the new technology is here to stay, they will willingly learn how to benefit from it or, at least, live with it.

Naysayers habitually oppose new technologies and often are very vocal about their opposition. They often gripe about any changes and will often never change if they don't have to or they quit before they are made to change "the way they do things."

The productivity vs. time curve will look different for each of these types of people. Think of how each person in your own organization fits into these four types. Think of how that impacts deriving the full benefits that you've carefully targeted. Think of how that impacts your ability to discover additional benefits once the technologies are implemented. Understanding the differences can help smooth out the rough spots during and after the implementation process.

LESSEN THE DISRUPTION; INCREASE THE EMPOWERMENT

Understanding the context in which processes exist, the democratizing potential of technology, and the types of people will help you achieve the goals stated above for a more rapid payoff from a smoother introduction of new technologies.

In addition, make the new technologies transparent to the user or, at least, make them as intuitive to operate as possible. Extra time in pre-planning the introduction of new technologies and training employees in the use of the technologies can provide a return many times greater than the hours spent in planning and training. You can achieve faster increases in productivity, reduced impact on customers, and lower burdens on support staff.

With proper planning and training, the productivity curve will increase at a faster rate and to a higher level than it might otherwise have achieved

Ed Mass is President of Mass Strategic Communications, Inc., a telecommunications consulting firm since 1993. They specialize in Transforming Telecommunications from a Tactical Tool To a Strategic Business Resource. We Integrate Business Strategies with Technology Opportunities.

We act as an extension of your staff. We are business strategists to increase the performance of your company through intelligent and cost effective use of technology.

Specifically, we consult on IP Telephone System Decisions, Service Provider Decisions for Voice and Data Services, and Services Audits to Inventory All Services and Discover Unused Services. We do all this within a framework of Vendor-Neutral Consulting.

CONCLUSION

IMPACT OF NEW TECHNOLOGIES BY 2030

According to the 2012 report, Global Trends 2030: Alternative Worlds, published the US National Intelligence Council, four technology arenas will shape global economic, social and military developments by 2030. They are information technologies, automation and manufacturing technologies, resource technologies, and health technologies.

INFORMATION TECHNOLOGIES

Three technological developments with an IT focus have the power to change the way we will live, do business and protect ourselves before 2030.

1. Solutions for storage and processing large quantities of data, including "big data", will provide increased opportunities for governments and commercial organizations to "know" their customers better. The technology is here but customers may object to

collection of so much data. In any event, these solutions will likely herald a coming economic boom in North America.

2. Social networking technologies help individual users to form online social networks with other users. They are becoming part of the fabric of online existence, as leading services integrate social functions into everything else an individual might do online. Social networks enable useful as well as dangerous communications across diverse user groups and geopolitical boundaries.

3. Smart cities are urban environments that leverage information technology-based solutions to maximize citizens' economic productivity and quality of life while minimizing resources consumption and environmental degradation.

AUTOMATION AND MANUFACTURING TECHNOLOGIES

As manufacturing has gone global in the last two decades, a global ecosystem of manufacturers, suppliers, and logistics companies has formed. New manufacturing and automation technologies have the potential to change work patterns in both the developed and developing worlds.

1. Robotics is today in use in a range of civil and military applications. Over 1.2 million industrial robots are already in daily operations round the world and there are increasing applications for non-industrial robots. The US military has thousands of robots in battlefields, home robots vacuum homes and cut lawns, and hospital robots patrol corridors and distribute supplies. Their use will increase in the coming years, and with enhanced cognitive capabilities, robotics could be hugely disruptive to the current global supply chain system and the traditional job allocations along supply chains.

2. 3D printing (additive manufacturing) technologies allow a machine to build an object by adding one layer of material at a time. 3D printing is already in use to make models from plastics in sectors such as consumers products and the automobile and aerospace industries. By 2030, 3D printing could replace some conventional mass production, particularly for short production runs or where mass customization has high value.

3. Autonomous vehicles are mostly in use today in the military and for specific tasks e.g. in the mining industry. By 2030, autonomous vehicles could transform military operations, conflict resolution, transportation and geo-prospecting, while simultaneously presenting novel security risks that could be difficult to address. At the consumer level, Google has been testing for the past few years a driverless car.

RESOURCE TECHNOLOGIES

Technological advances will be required to accommodate increasing demand for resources owing to global population growth and economic advances in today's underdeveloped countries. Such advances can affect the food, water and energy nexus by improving agricultural productivity through a broad range of technologies including precision farming and genetically modified crops for food and fuel. New resource technologies can also enhance water management through desalination and irrigation efficiency; and increase the availability of energy through enhanced oil and gas extraction and alternative energy sources such as solar and wind power, and biofuels. Widespread communication technologies will make the potential effect of these technologies on the environment, climate and health well known to the increasingly educated populations.

HEALTH TECHNOLOGIES

Two sets of health technologies are highlighted below.

1. Disease management will become more effective, more personalized and less costly through such new enabling technologies as diagnostic and pathogen-detection devices. For example, molecular diagnostic devices will provide rapid means of testing for both genetic and pathogenic diseases during surgeries. Readily available genetic testing will hasten disease diagnosis and help physicians decide on the optimal treatment for each patient. Advances in regenerative medicine almost certainly will parallel these developments in diagnostic and treatment protocols. Replacement organs such as kidneys and livers could be developed by 2030. These new disease management technologies will increase the longevity and quality of life of the world's ageing populations.

2. Human augmentation technologies, ranging from implants and prosthetic and powered exoskeleton to brains enhancements, could allow civilian and military people to work more effectively, and in environments that were previously inaccessible. Elderly people may benefit from powered exoskeletons that assist wearers with simple walking and lifting activities, improving the health and quality of life for aging populations. Progress in human augmentation technologies will likely face moral and ethical challenges.

CONCLUSION

The US National Intelligence Council report asserts that "a shift in the technological center of gravity from West to East, which has already begun, almost certainly will continue as the flows of companies, ideas, entrepreneurs, and capital from the developed world to the developing markets increase". I am not convinced that this shift will "almost certainly" happen. While the East, in particular Asia, will likely see the majority of technological applications, the current innovations are taking place mainly in the West. And I don't think it is a sure bet that the center of gravity for technological innovation will shift to the East.

www.ingramcontent.com/pod-product-compliance
Lightning Source LLC
Chambersburg PA
CBHW031423210526
45464CB00005B/2020